Practical Physics Workbook

for Secondary Schools

I0488935

Preface

This book is intended for use in Physics laboratories as a workbook for carrying out practical physics experiments by secondary school students and first year higher institution students. The objective is to have an all-in-one workbook from which various relevant physics experiments can be performed in a manner that also prepares students for practical physics examinations especially those of the West African Senior School Certificate Examination (WASSCE) and the National Examination Council (NECO).

We immensely acknowledge the West African Examination Council (WAEC) and the NECO whose examination papers were largely used in preparing this workbook. Plenty thanks to the numerous persons that contributed to make this book beautiful.

Experiment A1: Oscillatory Spiral Spring Experiment

Purpose: To determine the force constant of a given spiral spring by method of oscillations.

Apparatus: Retort stand, clamp and boss, a set of masses, a spiral spring, stop watch, split cork and mass hanger.

Procedure:

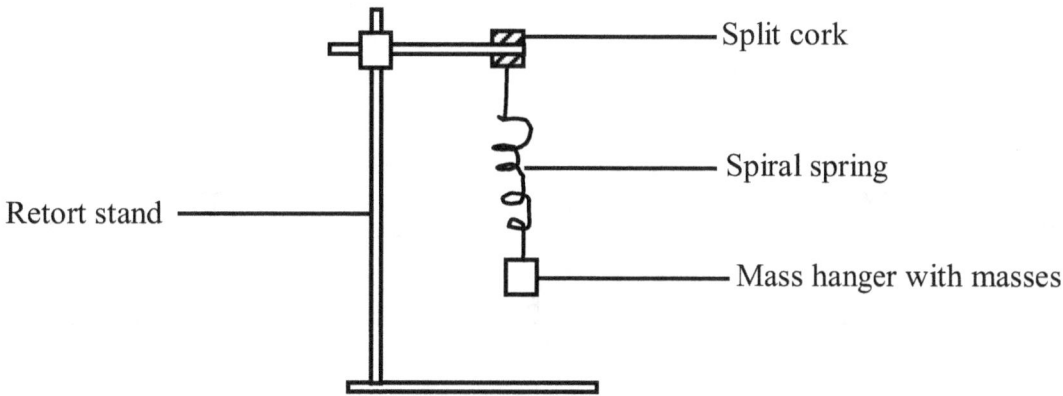

(a) You have been provided with a mass hanger and a set of masses, a retort stand, a spiral spring, clamp and boss, stop watch and split cork. Set up the system as shown in the diagram above by doing the following:

   (i)   Suspend the spiral spring vertically as shown in the diagram.
   (ii)   Suspend the mass hanger on the free end of the spiral spring and add a mass, m=50.0 g to the hanger.
   (iii)   Pull the hanger vertically downwards and release to set it into vertical oscillations.
   (iv)   Determine the time, t, for 20 complete oscillations.
   (v)   Evaluate the period, T, of the oscillation. And then evaluate T^2.

(vi)   Repeat the procedure for four other values of m=70, 90, 110 and 130 g. In each case, determine t and evaluate T and T^2. Tabulate your readings.

(vii) Plot a graph of T^2 on the vertical axis against m on the horizontal axis.

(viii) Determine the slope, s, of the graph and the intercept, I, on the vertical axis.

(ix)   Evaluate $k = \dfrac{4\pi^2}{s}$. [Take $\pi = \dfrac{22}{7}$].

(x)   State two precautions taken to ensure accurate results.

(b)

(i) Define period and frequency, with respect to a simple harmonic motion.

(ii) A force of magnitude 750 N is applied to the free end of a spiral spring of force constant $2.5 \times 10^5 Nm^{-1}$. Calculate the energy stored in the stretched spring.

Observations:

m (g)	t (s)	T (s)	T^2 (s^2)

_____

_____

_____

_____

_____

Experiment A2: Meter Rule Balancing Experiment

Purpose: To determine the mass of a meter rule by balancing on a knife edge (method of moments).

Apparatus: Two 100 g masses, meter rule, knife edge, thread and adhesive.

Procedure:

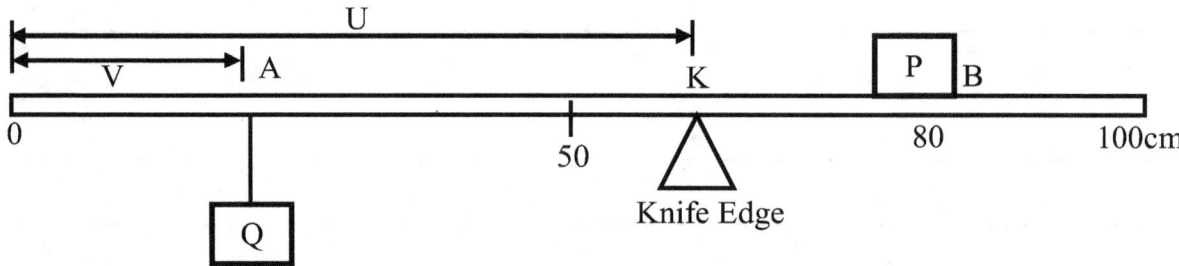

(a)

(i) Fix the 100 g mass, marked P, at B, the 80cm mark of the uniform meter rule, using an adhesive.

(ii) Suspend another 100 g mass marked Q at A, a distance $V = 10.0$ cm from the 0 cm mark of the meter rule.

(iii) Balance the whole arrangement horizontally on a knife edge as shown in the diagram above.

(iv) Measure and record the distance, U, of K from the 0 cm mark of the meter rule.

(v) Repeat the procedure for five other values of $V = 15.0, 25.0, 30.0$ and 35.0 cm.

(vi) In each case measure and record the corresponding values of U. Tabulate your readings.

(vii) Plot a graph of U on the vertical axis against V on the horizontal axis.

4

(viii) Determine the:
    I.   slope, s, of the graph
    II.   intercept, c, on the vertical axis

(ix) Evaluate:

    I. $k_1 = \left(\frac{1-2s}{s}\right) 100$
    II. $k_2 = \frac{2c}{s} - 160$

(x) State two precautions taken to ensure accurate results.

(b)

(i) State two conditions under which a rigid body at rest remains in equilibrium when acted   upon by three non-parallel coplanar forces.
(ii) Explain how the position of the center of gravity of a body affects the equilibrium of the body.

Observations:

V (cm)	U (cm)

_____

_____

Experiment A3: Hooke's Law Experiment

Purpose: To study Hooke's law using a spiral spring.

Apparatus: Retort stand, meter rule, spiral spring, split cork, pointer, a set of masses and mass hanger.

Procedure:

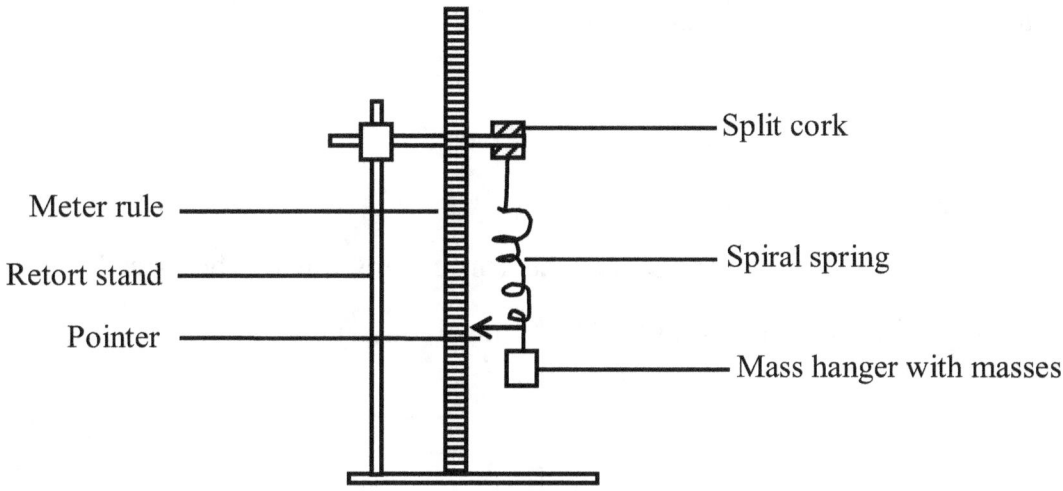

(a) You are provided with a meter rule, retort stand, spiral spring and a set of masses.

   (i) Suspend the given spiral spring vertically as shown in the diagram above.

   (ii) Attach the mass hanger and note the position of the pointer on the meter rule.

   (iii) Add a mass $m = 70$ g to the mass hanger and note the new position of the pointer. Determine the extension e produced.

7

(iv) Repeat the experiment for m = 90, 110, 130 and 150 g respectively. In each case determine the extension e produced. Ignore the mass of the mass hanger and tabulate your readings.

(v) Plot a graph with e on the vertical axis and m on the horizontal axis, starting both axes from the origin (0, 0).

(vi) Determine the slope of the graph and the intercept on the e axis. Also, determine the difference in the extension x when the mass is increased from 100 g to 150 g.

(vii) State **two** precautions taken to ensure accurate results.

(b)

      (i) State Hooke's law.

      (ii) Using your graph, deduce the force constant of the spiral spring you used in this experiment. (Take g as $10ms^{-2}$)

Observations:

m (g)	e (cm)

Experiment A4: Oscillatory Meter Rule Experiment

Purpose: To study the Young modulus of a meter rule using the oscillatory meter rule on a G-clamp.

Apparatus: Set of masses, meter rule, adhesive and G-Clamp.

Procedure:

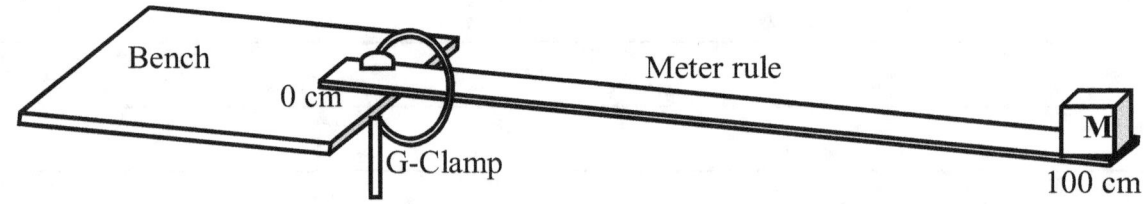

(a) You have been provided with a meter rule, a clamp and a set of masses.

  (i)   Clamp the meter rule to the edge of the bench such that 90 cm of the rule projects from the edge as shown in the diagram above. Ensure that the meter rule is capable of performing oscillatory motion.
  (ii) Fix a mass M=50 g at the free end of the rule.
  (iii) Deflect the rule slightly such that it performs vertical oscillation.
  (iv) Determine the time, t, for 10 complete oscillations.
  (v)   Calculate the period, T, of the oscillations and evaluate T^2.
  (vi) Repeat the procedure for four other values of M=100, 150, 200 and 250 g. In each case, determine and record the corresponding values of t, T and T^2. Tabulate your readings.
  (vii) Plot a graph of T^2 on the vertical axis against M on the horizontal axis, starting both axis from the origin (0,0).
  (viii) Determine the slope, s, of the graph, and its intercept, c, on the vertical axis.

10

(ix)   Evaluate $k = \frac{4\pi}{s}$.   [Take $\pi = \frac{22}{7}$].

(x)   From your graph determine the period T, when M=180g.

(xi)   State two precautions taken to ensure accurate results.

(b)

(i) Define Young modulus and force constant of an elastic material.

(ii)Explain simple harmonic motion

Observations:

M (g)	t (s)	T (s)	T^2 (s^2)

_____

_____

_____

_____

_____

_____

_____

_____

Experiment A5: Simple Pendulum Experiment

Purpose: To determine the value of acceleration due to gravity using a simple pendulum.

Apparatus: Pendulum bob, thread, retort stand, meter rule and clamp.

Procedure:

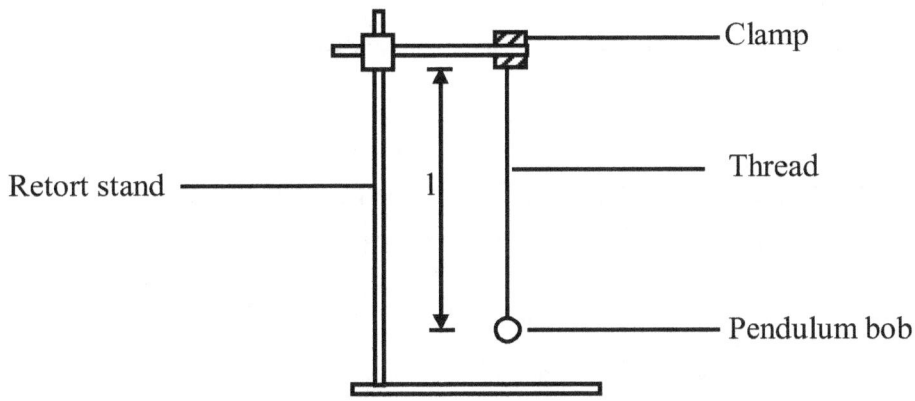

(a) You have been provided with a meter rule, retort stand, clamp, pendulum bob, and a piece of thread. Set up the system as in the diagram above, and perform the experiment as explained below:

(i) Measure and record the distance, l=130 cm from the center of the bob to the point of suspension of the pendulum.

(ii) Displace the pendulum through a small angle and release. Allow the pendulum to oscillate freely.

(iii) Determine the time, t, for 20 complete oscillations

(iv) Also determine the period, T, of the oscillations.

(v) Evaluate T^2 and L=l-30.

(vi) Repeat the procedure for four other values of l=110, 90, 70 and 50 cm.

(vii) In each case, determine t, and evaluate T, T^2 and L. Tabulate your readings.

(viii) Plot a graph of T^2 on the vertical axis against L on the horizontal axis, starting both axes from the origin (0,0).

(ix)   Determine the slope, s, of the graph, and the intercept, c, of the graph on the T^2 axis.

(x)   Evaluate:

   I. $k_1 = \dfrac{4\pi^2}{s}$ [Take $\pi = \dfrac{22}{7}$].

   II. $k_2 = \dfrac{c}{s}$.

(xi)   State two precautions taken to ensure accurate results.

(b)

   (i) State two factors that affect the period of a simple pendulum
   (ii) Explain the acceleration of free fall due to gravity.

Observations:

l (cm)	t (s)	T (s)	T^2 (s^2)	L (cm)

Experiment A6: Principle of Moments Experiment I

Purpose: To study the principle of moments.

Apparatus: Two known masses, meter rule, knife edge and thread.

Procedure:

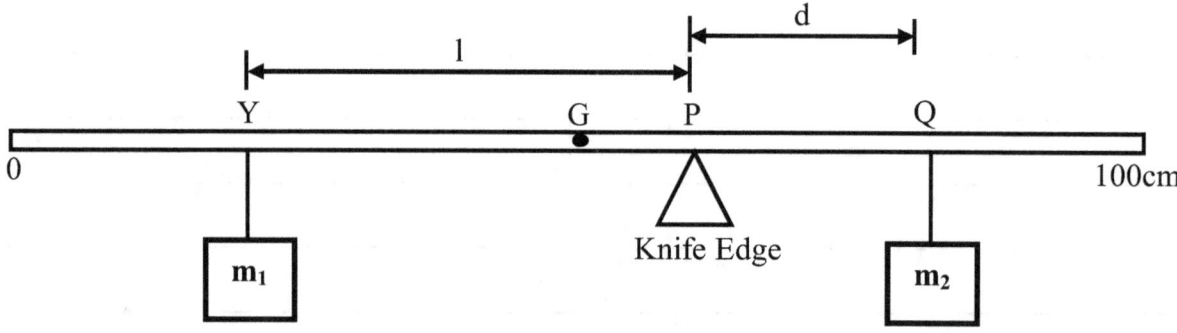

(a) You are provide with a meter rule, a knife edge, two pieces of thread and two masses m_1 and m_2.

(i) Record the values of m_1 and m_2.

(ii) Balance the meter rule horizontally on the knife edge and record the balance point G.

(iii) With the knife edge at the 60 cm mark of the meter rule, suspend m_1 at the 20 cm mark and m_2 at a suitable mark such that the rule balances horizontally as illustrated at the diagram above.

(iv) Record the position Y of m_1 and Q of m_2.

(v) Evaluate $l = P - Y$, and $d = Q - P$.

(vi) Repeat the procedure for four other positions of m_1 at the 18, 16, 14 and 12 cm marks.

16

(vii) In each case, evaluate and record l and d.

(viii) Tabulate your readings.

(ix) Plot a graph of l on the vertical axis against d on the horizontal axis.

(x) Determine the slope of the graph.

(xi) State two precautions taken to ensure accurate results.

(b)

(i) With the aid of a diagram, indicate the forces acting on the meter rule in the experimental set up above.

(ii) Define the moment of a force about a point and state it S. I. unit.

Observations:

$m_1 =$ _____ $m_2 =$ _____ $G =$ _____

Y (cm)	Q (cm)	l (cm)	d (cm)

_____

_____

_____

_____

Experiment A7: Principle of Moments Experiment II

Purpose: To determine the unknown mass of an sand in a measuring cylinder using the principle of moments.

Apparatus: Unknown mass of sand in a measuring cylinder, meter rule, knife edge and a known mass.

Procedure:

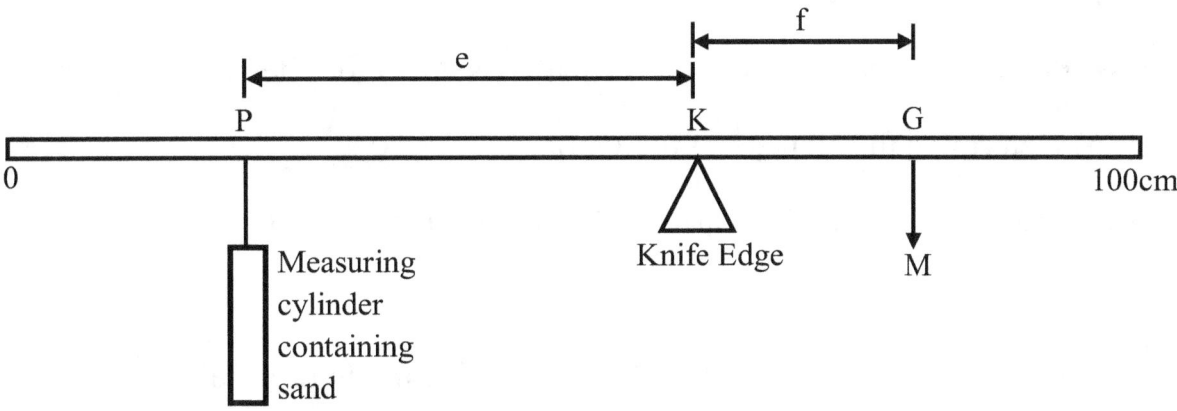

(a) You are provided with a uniform meter rule of mass, M indicated on its reverse side, a knife edge, a graduated measuring cylinder of known mass, m_1, marked on it and other necessary apparatus.

(i) Read and record the values of M and m_1.

(ii) Balance the meter rule horizontally on the knife edge. Read and record the balance point as G.

(iii) Tie a loop of thread round the neck of a measuring cylinder.

(iv) Fill the cylinder with the sand provided to the $2cm^3$ mark. Record the volume, V, of the sand.

19

(v) Hang the cylinder at the 2cm mark of the meter rule and adjust the center of the knife edge until the rule balances horizontally.

(vi) Read and record the new balance position K.

(vii) Determine the values of e and f.

(viii) Determine the mass, m_2 of the sand in the measuring cylinder.

Hint: $m_2 = \left(\frac{M \times f}{e}\right) - m_1$

(ix) Repeat the procedure by filling the measuring cylinder to the mark V = 4, 6, 8 and 10 cm^3. In each case ensure that the measuring cylinder is kept constant at the 2cm mark on the meter rule.

(x) Tabulate your readings.

(xi) Plot a graph with m_2 on the vertical axis and V on the horizontal axis.

(xii) Determine the slope, s of the graph.

(xiii) State two precautions taken to ensure accurate results.

(b)

(i) Determine the mass of $7.5cm^3$ of the sand using your graph.

(ii) A gold coin of mass 350.0 g has uniform cross sectional area of $25.0\ cm^2$. Calculate its thickness. [$Density\ of\ gold = 19.3\ gcm^{-3}$].

Observations:

M = _____ m_1 = _____   G = _____

V (cm³)	e (cm)	f (cm)	m_2 (g)

_____
_____
_____
_____
_____
_____
_____
_____
_____
_____
_____
_____
_____
_____
_____
_____
_____

Experiment A8: Archimedes' Principle and Moments

Purpose: To jointly study the principle of Archimedes with those of moments.

Apparatus: A 100g mass and an unknown mass labeled m, drilled meter rule, beaker containing water, knife edge and thread.

Procedure:

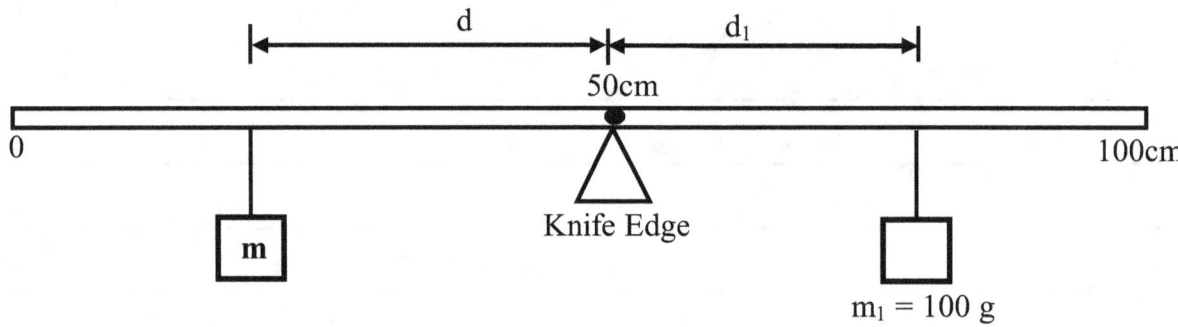

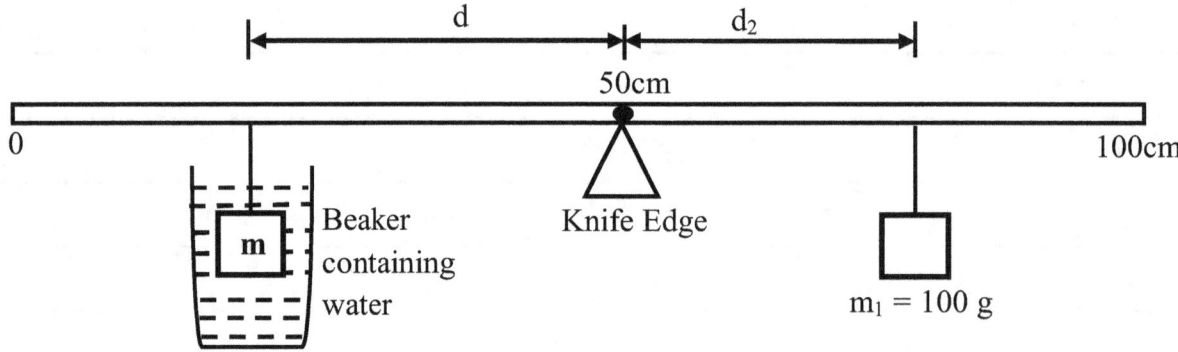

(a) You have been provided with a drilled meter rule and other apparatus needed for this experiment. Carry out the experiment as explained below:

(i) Pivot the meter rule at the 50cm mark. This should remain the same throughout the experiment.

(ii) Suspend the object marked **m** at the 10cm mark of the meter rule. On the other side of the pivot, suspend the known mass $m_1 = 100g$ and adjust its position until the rule balances horizontally.

(iii) Read and record the distances **d** and d_1 of **m** and m_1, respectively from the pivot.

(iv) Repeat the experiment with **m** suspended at the 15, 20, 25, and 30cm marks respectively.

(v) In each case, adjust the position of m_1 and determine **d** and d_1.

(vi) Also, repeat the entire experiment with **m** now completely immersed in a beaker of water and at 10, 15, 20, 25 and 30cm marks respectively. In each case, read and record the new distance d_2, of mass m_1 from the pivot.

(vii) Evaluate $d_1 - d_2$. Tabulate your values of **d**, d_1, d_2 and $d_1 - d_2$.

(viii) Plot a graph of d_1 on the vertical axis and $d_1 - d_2$ on the horizontal axis.

(ix) Determine the slope **s** of the graph.

(x) State **two** precautions taken to ensure accurate results.

(b)

(i) State Archimedes' principle.
(ii) An object weighs 5.4 N in air and 3.2 N when completely immersed in water. Calculate the relative density of the object.

Observations:

Position of m (cm)	d (cm)	d_1 (cm)	d_2 (cm)	d_1-d_2 (cm)

_____

_____

_____

_____

_____

_____

_____

_____

_____

_____

_____

_____

_____

_____

Experiment A9: Compound Pendulum Experiment

Purpose: To study properties of a compound pendulum, and to use it to determine the value of acceleration due to gravity at the location of the experiment.

Apparatus: Retort stand, drilled meter rule, knife-edge, cork and steel needle.

Procedure:

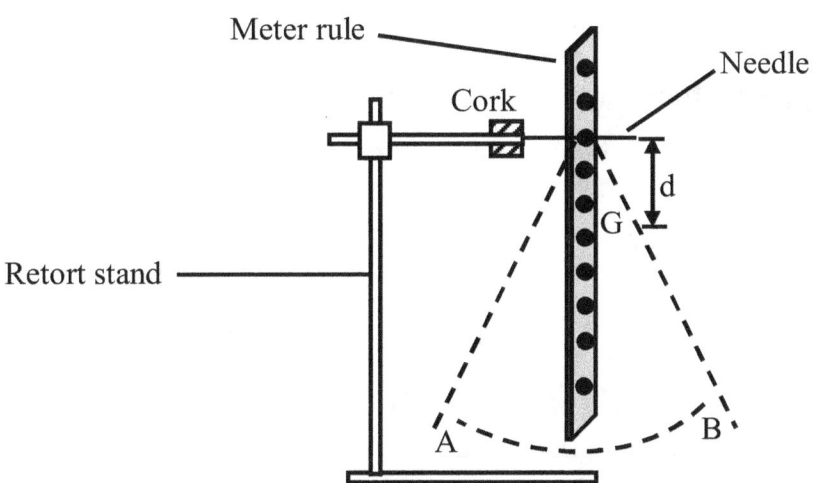

(a) You have been provided with a drilled meter rule, retort stand, knife-edge, cork and steel needle. Perform the experiment as detailed below:

(i) Use the knife-edge provided to determine and record the centre of gravity G of the meter rule.

(ii) Push the steel needle firmly into the cork, then clamp the cork on the retort stand horizontal.

(iii) Measure and record the distance **d** between **G** and the small hole at the 5cm mark on the meter rule. Through this hole, suspend the meter rule freely on the needle as shown in the diagram above.

(iv) Slightly displace the free end of the meter rule sideways and release to set it into oscillations in the vertical plane.

(v) Determine the time **t** for 20 complete oscillations of the rule, and calculate the period **T** of oscillation.

(vi) Evaluate $\mathbf{d^2}$ and $\mathbf{T^2d}$.

(vii) Repeat the experiment for the holes at the 10, 15, 20, and 25 cm marks on the meter rule respectively.

(viii) In each case, determine the values of **t, T, $\mathbf{d^2}$** and $\mathbf{T^2d}$. Tabulate your readings.

(ix) Plot a graph of $\mathbf{T^2d}$ on the vertical axis against $\mathbf{d^2}$ on the horizontal axis, starting both axes from the origin (0,0).

(x) Determine the slope *s* of the graph and its intercept *I* on the vertical axis

(xi) Evaluate

    I. $C = \dfrac{4\pi^2}{s}$

    II. $K = \sqrt{\dfrac{I}{s}}$

(xii) State two precautions taken to ensure accurate results.

(b)

    (i) Define the centre of gravity of an object and state how it is related to the stability of the object.

    (ii) State two reasons why the acceleration of free fall due to gravity varies from place to place on the earth's surface.

Observations:

G = _____

Position of holes (cm)	d (cm)	t (s)	T (s)	d^2 (cm^2)	T^2d (cms^2)

_____

_____

_____

_____

_____

_____

_____

_____

_____

_____

_____

_____

_____

Experiment A10: Equilibrium and Pulley Experiment

Purpose: To study the principles of equilibrium using a system of pulleys.

Apparatus: Two pulleys, set of masses, meter rule, force board, paper and thread.

Procedure:

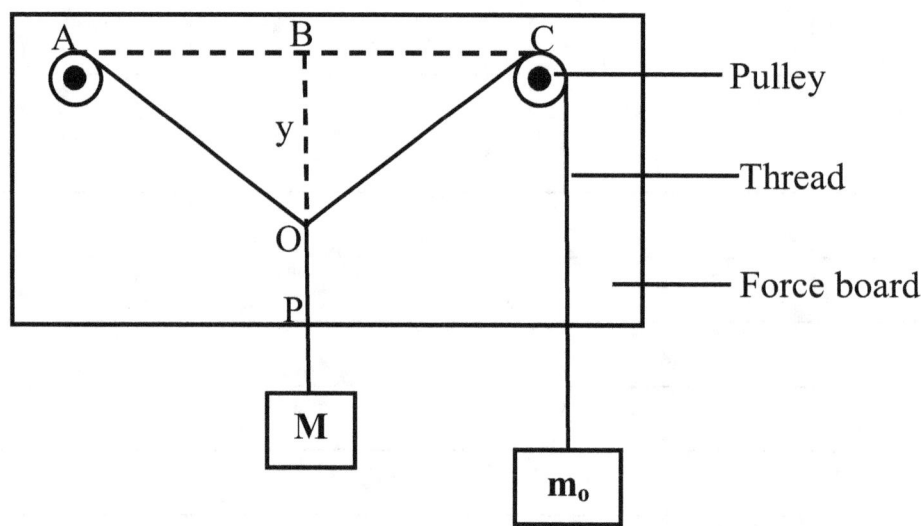

(a) In the diagram above, a thread **AC**, fixed at pulley **A** passes over pulley **C** on a force board and carries an unknown mass m_o. Retain this mass m_o throughout the experiment.

> (i) Draw a line along the direction of **AC** on the paper held behind the thread.
> (ii) Locate the midpoint **B** of **AC** and mark its position on this line. Draw **BP** at right angles to **AC**.

(iii) By means of a loop of thread, suspend a mass **M** = 50 g from **AC** and adjust the position of the loop so that the line of action of the weight of **M** lies along **BP**. Ensure that **M** and m_o hang off the force board.

(iv) Measure **BO** = **y** and **AO**. Evaluate **y/AO.**

(v) Repeat the experiment for **M** = 70, 90, 110 and 130 g respectively.

(vi) In each case, determine the corresponding values of **y**, **AO** and **y/AO.** Tabulate your readings.

(vii) Plot a graph of **y/AO** on the vertical axis and **M** on the horizontal axis.

(viii) Determine the slope **s** of the graph.

(ix) State **two** precautions taken to ensure accurate results.

(Attach your traces to you answer script)

(b)

(i) Distinguish between the **resultant** and the **equilibrant** of forces

(ii) State **two** conditions necessary for the equilibrium of three non-parallel coplanar forces.

Observations:

M (g)	y (cm)	AO (cm)	y/AO

_____

_____

_____

Experiment B1: Glass Block Experiment I

Purpose: To study Snell's law of refraction using a rectangular glass block.

Apparatus: Rectangular glass block, 4 optical pins, drawing board, paper, rule and compass.

Procedure:

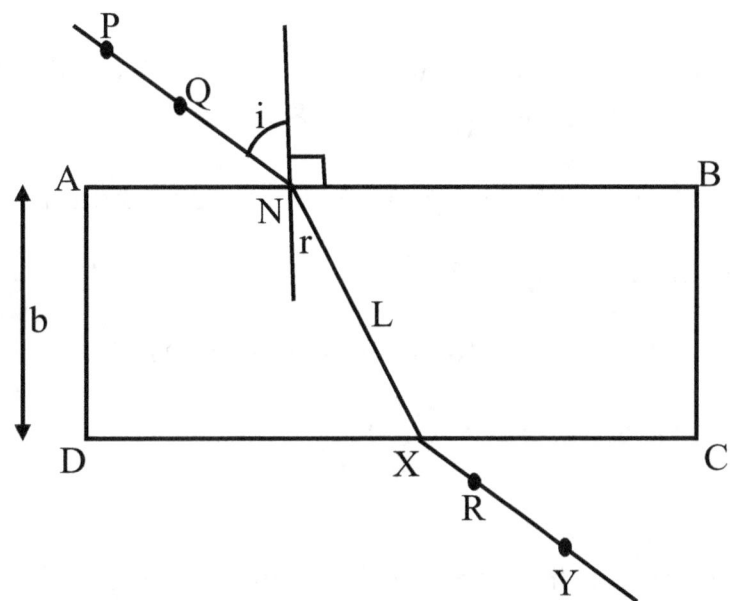

(a) You have been provided with a rectangular glass block and other apparatus require for this experiment. Carry out the experiment as described below:

  (i) Measure and record the thickness **b** of the glass block provided.
  (ii) Trace the outline ABCD of the glass block on the sheet of paper as shown in the diagram above.

(iii) Remove the block and draw a normal at **N.**

(iv) Draw an incident ray such that the angle of incidence $i = 25^0$.

(v) Fix **two** pins at points **P** and **Q** on the incident ray. Replace the glass block and fix **two** other pins at points **R** and **Y** such that the pins appear to be in a straight line with the images of the pins at **P** and **Q** when viewed through the side **DC** of the glass block.

(vi) Remove the block and join the points at **R** and **Y**, producing the line to meet **DC** and **X.**

(vii) Join **NX** and measure its length **L**. Evaluate L^{-2} and $\sin^2 i$.

(viii) Repeat the experiment for $i = 35^0, 45^0, 55^0$ and 65^0.

(ix) In each case, determine the corresponding values of **L**, L^{-2} and $\sin^2 i$. Tabulate your readings.

(x) Plot a graph of L^{-2} on the vertical axis and $\sin^2 i$ on the horizontal axis starting both axes from the origin (0,0).

(xi) Determine the slope **s** of the graph and the intercept **I** on the vertical axis.

(xii) Evaluate the expression $K = (I/s)^{\frac{1}{2}}$.

(xiii) State **two** precautions taken to ensure accurate results.

(Attach your traces to you answer script).

(b)

(i) State Snell's law of refraction and explain why refraction occurs at the boundary between two media.

(ii) Using your graph, deduce the value **L** when $I = 0^0$.

Observations:

b = _____

i (^{o})	L (cm)	L^{-2} (cm^{-2})	sin^{2}i

_____

_____

_____

_____

_____

_____

_____

_____

_____

_____

_____

_____

_____

_____

_____

Experiment B2: Plane Mirror Experiment

Purpose: To study the laws of reflection using plane mirrors.

Apparatus: Plane mirror, 4 optical pins, drawing board, paper, rule and compass.

Procedure:

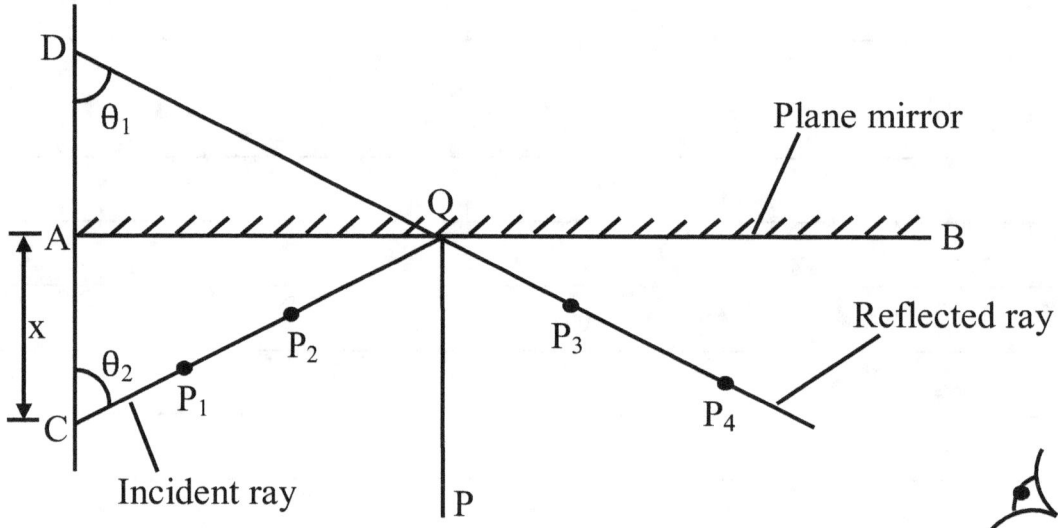

(a) Using the diagram above as a guide, carry out the following instructions:

(i) Fix the drawing paper provided on the drawing board.

(ii) Place the mirror vertically with its longer side resting on the drawing paper.

(iii) Trace the outline AB of the mirror. Remove the mirror.

(iv) Draw a normal PQ to meet the outline at the middle Q.

(v) Draw a straight line through A to meet the outline of the mirror at right angle.

(vi) Trace the incident ray, CQ with pins P_1 and P_2 so that it meets the perpendicular line through A at C, and such that CA = x = 1.0 cm.

(vii) Replace the mirror on its outline. Locate the images of P_1 and P_2 using two other pins P_3 and P_4 so that P_3 and P_4 and the images of P_1 and P_2 are in a straight line.

(viii) Remove the mirror and pins P_3 and P_4. Draw a straight line through the pin points to meet AB at Q and CA produced at D.

(ix) Measure and record angle ACQ as θ_1 and angle ADQ as θ_2. Also record the value of x. Evaluate $\theta = \frac{1}{2}(\theta_1 + \theta_2)$, x^{-1} and $\tan \theta$.

(x) Repeat the procedures for four other values of x= 2.0, 3.0, 4.0 and 5.0 cm. Tabulate your readings.

(xi) Plot a graph of $\tan \theta$ on the vertical axis against x^{-1} on the horizontal axis.

(xii) Determine the slope, s, of the graph. Evaluate k=2s.

(xiii) State two precautions taken to ensure accurate results.

[Attach your traces to your answer booklet]

(b)

(i) State the laws of reflection of light.
(ii) An object is situated 25cm in front of a plain mirror. Determine the distance of the image from the object. What is the size of the image relative to the object?

Observations:

x (cm)	θ_1 (°)	θ_2 (°)	θ (°)	x^{-1} (cm^{-1})	$\tan \theta$

_____
_____
_____
_____
_____
_____
_____
_____
_____
_____
_____
_____
_____
_____
_____
_____

Experiment B3: Cooling Curve Experiment

Purpose: To study factors that affect heat losses from a material using heated water in different containers.

Apparatus: Water, two tins made of materials with different heat conductivities, bunsen burner, measuring cylinder, thermometer, clock/watch and wooden stand.

Procedure:

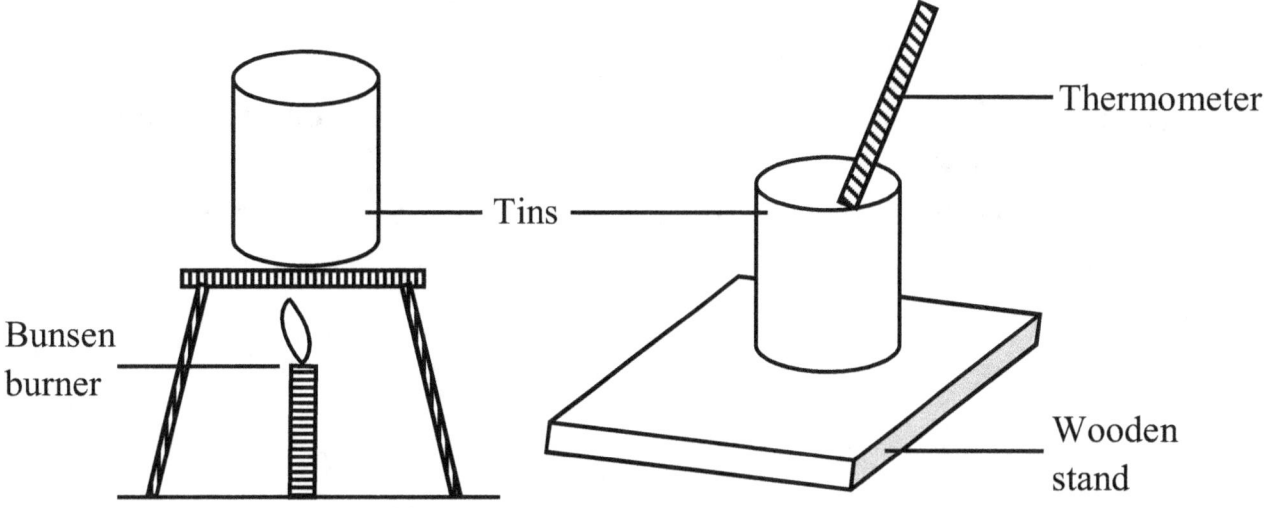

(a)

(i) You are provided with a measuring cylinder, two different tins labeled C and D, thermometer and other necessary materials.

(ii) Use the measuring cylinder provided to measure $100 \, cm^3$ of water and pour it into the tin labeled C.

(iii) Heat the water in the tin almost to boiling $(90^0 C)$.

(iv) Remove the tin and place it on the wooden stand provided.

(v) Insert the thermometer into the tin and record the temperature of water every minute starting from 60^0C.

(vi) Repeat the experiment with the tin labeled D using exactly the same volume of water and temperature range. Tabulate your readings.

(vii) On the same graph sheet and using the same axes and scales, plot two graphs of temperature on the vertical axis and time on the horizontal axis from the readings obtained using tins C and D.

(viii) Label the graph appropriately as C and D to correspond with the tins used.

(ix) From each graph, read off the time taken to cool from 85^0C to 65^0C.

(x) State two precautions taken to ensure accurate results.

(b)

    (i) Explain how heat losses by radiation and convection are minimized in a vacuum flask.

    (ii) State four factors which affect the rate of evaporation of a liquid in an open container.

Observations:

Time (minutes)	C temperature (°C)	D temperature (°C)

Experiment B4: Glass Block Experiment II

Purpose: To determine the refractive index of a rectangular glass block using Snell's law and the principles of refraction of light.

Apparatus: Rectangular glass block, 4 optical pins, paper, drawing board, rule and protractor.

Procedure:

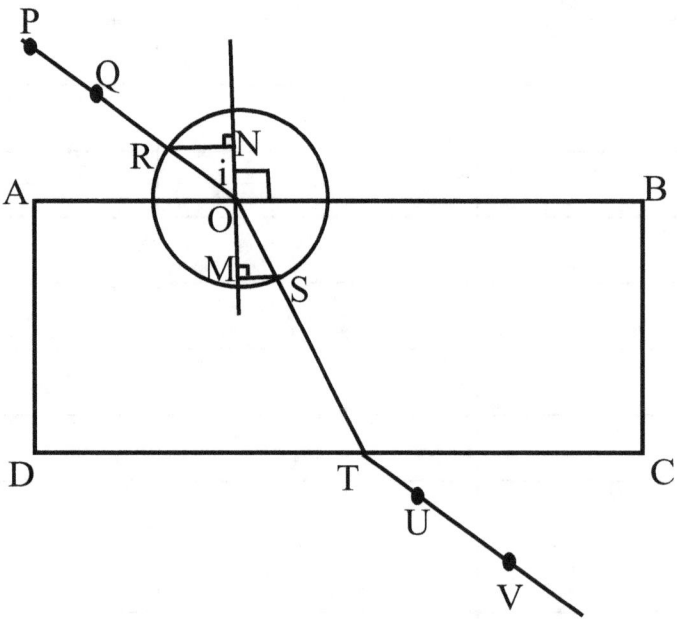

(a) Use the apparatus provided to carry out the following experiment.

   (i) Trace the outline **ABCD** of the glass block on the sheet of paper as shown above.

   (ii) Remove the block and draw the normal at **O.**

   (iii) Draw an incident ray such that the angle of incidence **i** = 30^0. Fix two pins at point **P** and **Q** on the incident ray.

(iv) Replace the block and fix two other pins at points **U** and **V** such that the pins appear to be in a straight line with the images of the pins at **P** and **Q** when viewed through the block.

(v) Remove the block and join the points at **V** and **U**, producing the line to meet **DC** at **T**. Join **OT**.

(vi) With **O** as centre and using any convenient radius, draw a circle to cut the incident and refracted rays at **R** and **S** respectively.

(vii) Draw the perpendiculars **RN** and **MS**. Measure and record **RN** and **MS.**

(viii) Repeat the experiment for **i** = 40^0, 50^0, 60^0 and 70^0 respectively.

(ix) In each case, determine and record the corresponding values of **RN** and **MS.** Tabulate your readings.

(x) Plot a graph **RN** on the vertical axis and **MS** on the horizontal axis.

(xi) Determine the slope **s** of the graph.

(xii) State **two** precautions taken to ensure accurate results.

(Attach your traces to your answer script)

(b)

    (i) Explain refraction.
    (ii) Draw a diagram showing why a meter rule, partly immersed in water and placed obliquely to the surface, appears bent at the surface.

Observations:

i (°)	RN (cm)	MS (cm)

_____

_____

_____

_____

_____

_____

_____

_____

_____

_____

_____

_____

_____

_____

_____

Experiment B5: Triangular Glass Prism Experiment

Purpose: To study refraction of light through a triangular glass prism.

Apparatus: Triangular glass prism, 4 optical pins, drawing board, paper, rule and compass.

Procedure:

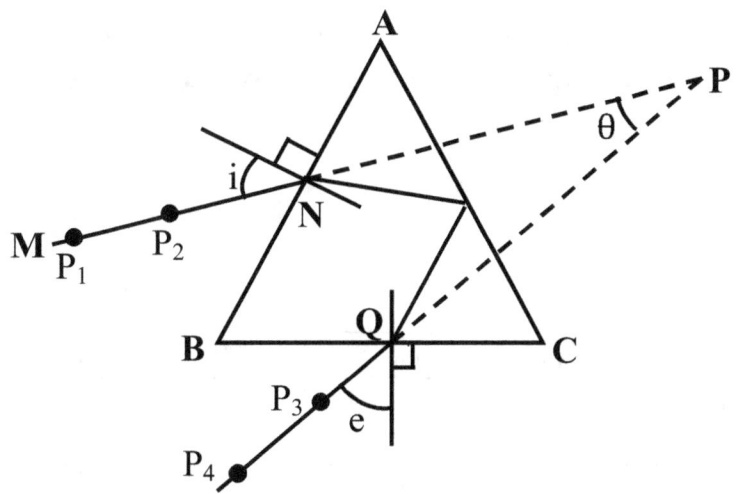

(a) Using the diagram above as a guide:

    (i)   Trace the outline ABC of the equilateral triangular glass prism provided.

    (ii)   Remove the prism. Draw a line MN such that it makes an angle $i = 5°$ with the normal at N on side AB of the outline.

    (iii)   Fix two pins at P_1 and P_2 on MN. Replace the prism on its outline.

    (iv)   Looking through the face BC of the prism, fix one pin at P_3 and another at P_4 such that they are in a straight line with the images of the pins at P_1 and P_2.

    (v)   Remove the prism and the pins. Draw a line to join P_4 and P_3. Produce line P_4P_3 to meet the line BC of the outline at Q and line MN produced at P.

(vi) Draw a normal to BC at Q. Measure and record the angles θ and e. Evaluate $\Phi = i + e$.

(vii) Repeat the procedure, using different outline in each case for four other values of $i = 10°$, $15°$, $20°$ and $25°$ respectively. Evaluate $\Phi = i + e$ in each case. Tabulate your readings.

(viii) Plot a graph of θ on the vertical axis against Φ on the horizontal axis starting both axes from the origin (0,0).

(ix) Determine the slope of the graph and the intercept on the vertical axis.

(x) State two precautions taken to ensure accurate results.

(Attach your traces to your answer booklet)

(b)

(i) Explain what is meant by the statement *the refractive index of glass is 1.5*.

(ii) Calculate the critical angle of a medium of refractive index 1.65 when light passes from the medium to air.

Observations:

i (°)	θ (°)	e (°)	Φ (°)

_____

_____

_____

Experiment B6: Water Refraction Experiment

Purpose: To study refraction in water.

Apparatus: Water, measuring cylinder, 2 optical pins, retort stand and rule.

Procedure:

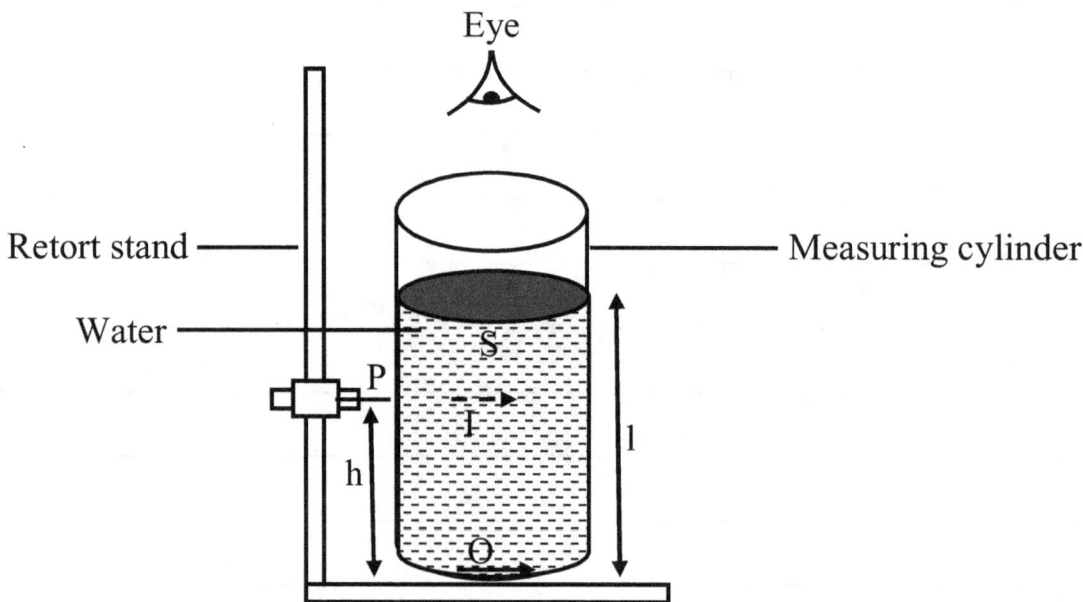

(a) Using the above diagram, carry out the experiment as described below:

(i) Place the pin O, horizontally inside the cylinder provided. Pour some water on the pin in the cylinder   such that the length of the water column, $l = SO = 10.0$ cm, where S represents the water meniscus.

(ii) Insert another pin, P, in the cork held by the boss of the retort stand.

(iii) Adjust the position of P vertically upward or downward until it coincides with the image I of O formed by refraction at S.

(iv) Read and record the distance $h = PO$.

(v) Repeat the procedure for four other values of l = 15, 20, 25 and 30cm.

(vi) In each case measure and record the corresponding values of h. Tabulate your readings.

(vii) Plot a graph of h on the vertical axis against l on the horizontal axis.

(viii) Determine the slope, s, of the graph.

(b)

    (i) Explain total internal reflection of light.

    (ii) A rectangular glass prism of thickness 10 cm and refractive index 1.6 is placed on the page of a book. The prints on the book are viewed vertically downward from the above. Determine the apparent upward displacement of the prints.

Observations:

l (cm)	h (cm)

_____

_____

_____

_____

_____

Experiment B7: Glass Block and Plane Mirror Experiment

Purpose: To study a combination of refraction and refraction of light through a rectangular glass block and on a plane mirror respectively.

Apparatus: Rectangular glass block, plane mirror, 4 optical pins, drawing board, paper, rule and compass.

Procedure:

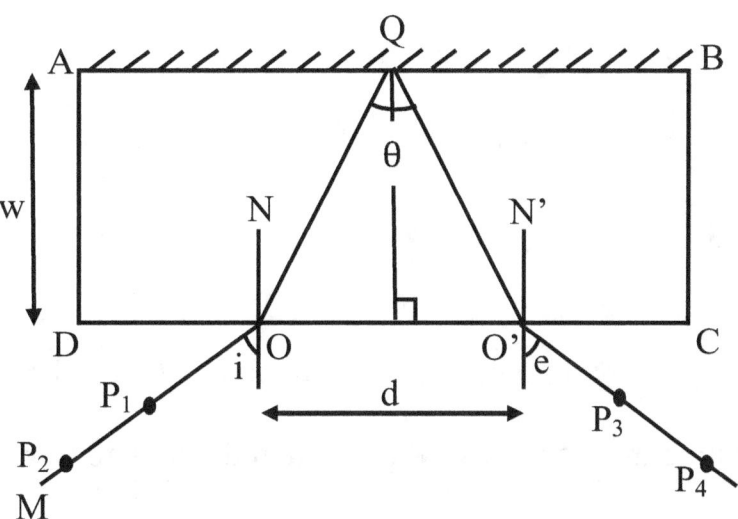

(a) You are provided with a glass block, plain mirror and optical pins.

   (i) Place the glass block on the paper provided, and trace its outline ABCD as shown in the diagram above.

   (ii) Remove the block, measure and record the width w of the block.

   (iii) Draw a normal ON to DC at a point about one quarter the length of DC.

   (iv) Draw a line making an angle $i = 10^0$ with the normal.

(v) Replace the block on its outline and mount the plain mirror vertically behind the block such that it makes good contact with the face AB.

(vi) Stick two pins P_1 and P_2 on the line MO.

(vii) Looking through the face CD, stick two other pins P_3 and P_4 such that they appear to be in straight line with the images of P_1 and P_2 seen through the block.

(viii) Join P_3 and P_4 with a straight line and extend it to touch the face DC at O'.

(ix) Draw a perpendicular line from the midpoint of OO' to meet AB at Q.

(x) Draw lines OQ, O'Q and normal O'N' produced.

(xi) Measure and record θ, e and d.

(xii) Evaluate $m = \sin e$ and $n = \cos \left(\dfrac{\theta}{2} \right)$.

(xiii) Repeat the procedure for $i = 20^0$, 30^0, 40^0 and 50^0. Tabulate your readings.

(xiv) Plot a graph with m on the vertical axis and n on the horizontal axis.

(xv) Determine the slope, s, of the graph and evaluate $\Phi = 2ws$.

(xvi) State two precautions taken to ensure accurate results.

(b)

    (i) Explain the term refractive index and give a mathematical expression for it in terms of wavelength.

    (ii) State the conditions necessary for total internal reflection to occur for a given pair of media.

Observations:

w = _____

i (°)	θ (°)	e (°)	d (cm)	m	N

_____

_____

_____

_____

_____

_____

_____

_____

_____

_____

_____

_____

_____

_____

Experiment B8: Heat Capacity Experiment

Purpose: To determine the specific heat capacity of a metallic bob.

Apparatus: Metallic bob, thread, copper calorimeter, beaker containing water, tin, thermometer, wooden stand and bunsen burner.

Procedure:

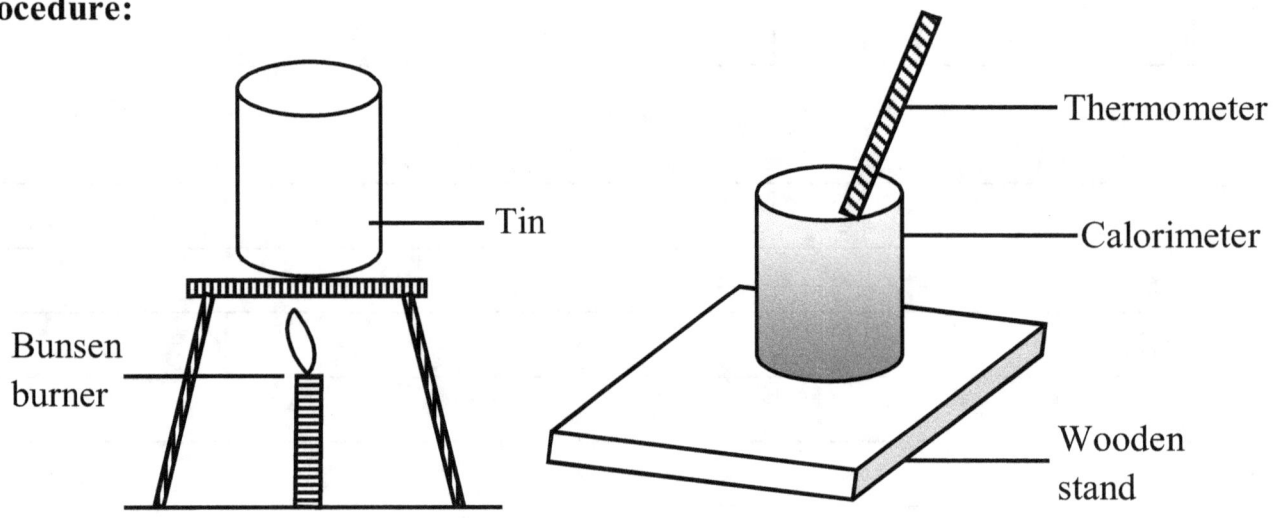

(a) Carry out the experiment as explained below:

(i) Using the provided weighing balance, measure and record the mass of the empty calorimeter, m_c and the mass of the metallic bob, m_b.

(ii) Pour some quantity of the provided water at room temperature into the calorimeter to about ¾ full. Using the weighing balance, measure and record the mass, m_{wc} of the calorimeter and water contained in it.

(iii) Also, using the thermometer, read and record the initial temperature, t_{wi} of water in the calorimeter.

(iv) Put some water into the tin. Using a thread tied to the metallic bob, suspend the bob in the tin containing water so that the bob is completely immersed in the water. Use the thermometer to read and record the initial temperature, t_{bi} of the water in the tin where the bob is immersed.

(iv) Place the tin of water and bob on the Bunsen burner and heat until the water is about to start boiling. Stop the heating and remove the tin and its content away from the burner.

(v) Use the thermometer to record the final temperature, t_{bf} of the heated water in the tin containing the bob.

(vi) Using the thread tied to the bob, quickly transfer the bob from the tin to the calorimeter containing water at room temperature. Close the calorimeter and stir the mixture.

(vii) Use the thermometer to read and record the final temperature, t_f of water in the calorimeter containing the bob.

(viii) Evaluate:

$$c_b = \frac{[(m_{wc} - m_c) \times c_w \times m_c \times c_c] \times (t_f - t_{wi})}{m_b \times (t_f - t_{bi})}$$

[Take the specific heat capacity of the copper calorimeter as $c_c = 400 \ JKg^{-1}K^{-1}$ and specific heat capacity of water as $c_w = 4200 \ JKg^{-1}K^{-1}$].

(ix) State two precautions taken to ensure accurate results.

(b)

(i) Define specific heat capacity.

(ii) How much heat is lost by a 500 g copper ball in cooling from 75°C to 25°C?

[Take the specific heat capacity of copper as $400 \ JKg^{-1}K^{-1}$]

Observations:

$m_{wc} =$ _____ $m_c =$ _____ $m_b =$ _____

$t_{wi} =$ _____ $t_{bi} =$ _____ $t_f =$ _____

_____

_____

_____

_____

_____

_____

_____

_____

_____

_____

_____

_____

_____

_____

_____

_____

_____

_____

_____

_____

Experiment B9: Glass Prism and Plane Mirror Experiment

Purpose: To study a combination of refraction and reflection through a triangular glass prism and a plane mirror respectively.

Apparatus: Triangular glass prism, plane mirror, 4 optical pins, drawing board, paper, rule and compass.

Procedure:

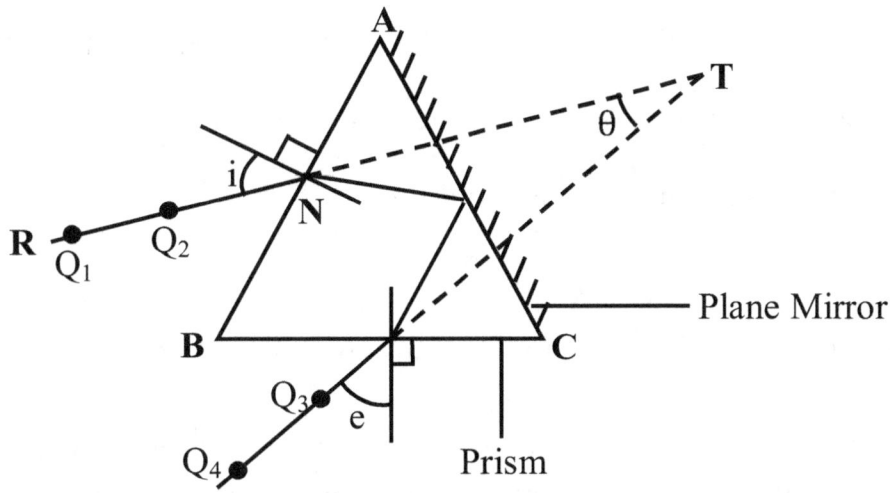

(a) Carry out the experiment as explained below:

   (i) Trace the outline **ABC** of the equilateral triangular glass prism as shown above. Remove the prism.

   (ii) Draw a line **RN** such that it makes an angle $i = 20^0$ with the normal. Erect two pins at Q_1 and Q_2 on the line **RN**.

(iii) Replace the prism. Place the reflecting surface of the plane mirror in contact with the face **AC** of the prism.

(iv) Looking through the face **BC** of the prism, fix one pin at **Q₃** and another pin at **Q₄** such that they appear to be in a straight line with the images of the pins at **Q₁** and **Q₂**.

(v) Remove the prism, the mirror and the pins. Draw a line to join points **Q₄** and **Q₃**. Produce the line **Q₄ Q₃** and that of **RN** to meet at **T**.

(vi) Measure and record angles e and θ.

(vii) Repeat the experiment for **i** = 25^0, 30^0, 40^0 and 50^0 respectively, using a different outline in each case. Determine and record the corresponding values of **e** and θ for each tracing. Tabulate your readings.

(viii) Plot a graph of θ on the vertical axis and **e** on the horizontal axis, starting both axes from the origin (0,0).

(ix) Determine the slope of the graph and the intercept on the vertical axis.

(x) State **two** precautions taken to ensure accurate results.

(Attach your tracings to your answer booklet).

(b)

(i) Draw a ray diagram showing how a right-angled isosceles glass prism may be used to invert a beam of light.
(ii) With the plane mirror removed in the experiment above, a ray of light is incident normally on the face **AB** of the prism. Draw a labeled ray diagram showing the path of the ray as it passes through and out of the prism.

Observations:

i (°)	e (°)	θ (°)

_____

_____

_____

_____

_____

_____

_____

_____

_____

_____

_____

_____

_____

_____

Experiment B10: Converging Lens Experiment

Purpose: To study refraction through a converging lens, and to determine the focal length of the lens.

Apparatus: Ray box with cross wire, converging lens, screen, meter rule and plane mirror.

Procedure:

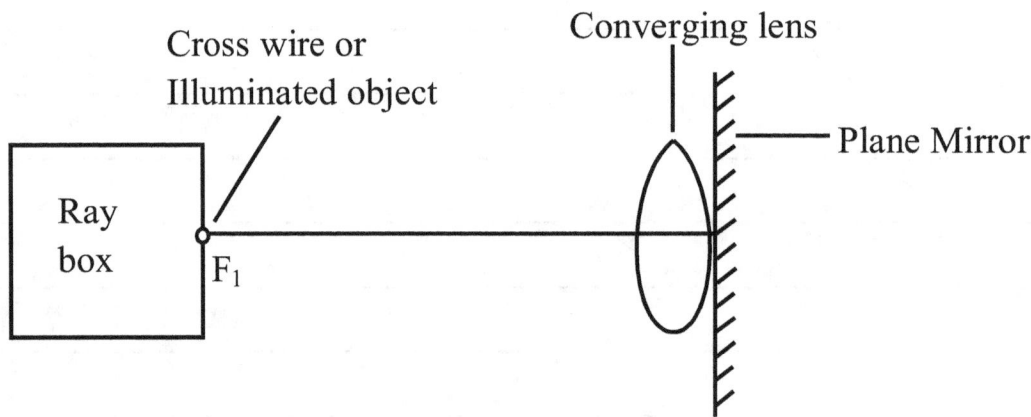

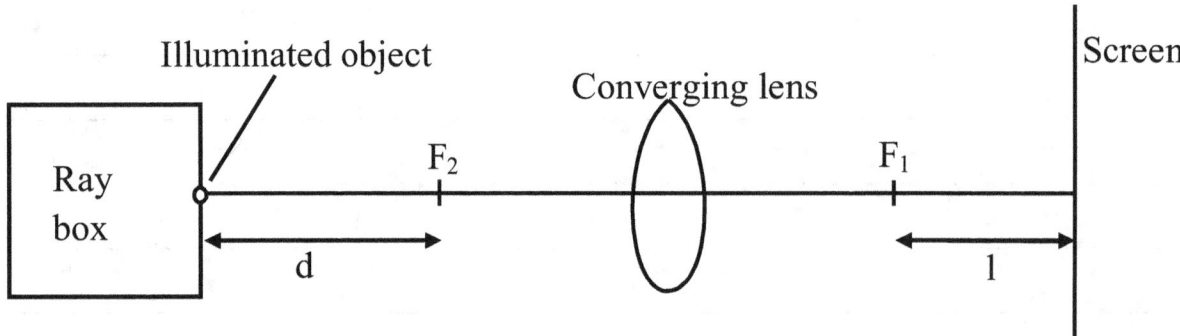

(a) You are provided with a converging lens, a plane mirror, a ray box, a screen and a meter rule.

(i) Mount the plane mirror behind the lens and place the ray box in front of it as shown in the first diagram above.

(ii) Adjust the position of the ray box until a sharp image of the cross wire is obtained in front of the box, beside the cross wire.

(iii) Measure and record the distance of the cross wire from the lens. This position of the cross wire is F_1.

(iv) Repeat the procedure with the lens reversed to obtain F_2.

(v) Replace the mirror with a screen.

(vi) Move the ray box a distance d = 10.0 cm from F_2 and adjust the position of the screen beyond F_1 until a sharp image of the wire is formed on the screen as shown in the second diagram.

(vii) Measure and record the distance, l between F_1 and the screen. Evaluate l^{-1}.

(viii) Repeat the procedures (v), (vi) and (vii) for four other values of d =15.0, 20.0, 25.0 and 30.0 cm. Tabulate your readings.

(ix) Plot a graph of l^{-1} on the vertical axis against d on the horizontal axis.

(x) Determine the slope, s, of the graph.

(xi) Evaluate $\frac{1}{\sqrt{s}}$.

(xii) State two precautions taken to ensure accurate results.

(b)

(i) Define the principal focus of a converging lens.
(ii) An object and its real image are located at distances 25.0 cm and 40.0 cm from the two principal foci of a converging lens respectively. Calculate the focal length of the lens.

Observations:

$F_1 =$ _____ $F_2 =$ _____

d (cm)	l (cm)	l^{-1} (cm^{-1})

_____

_____

_____

_____

_____

_____

_____

_____

_____

_____

_____

_____

_____

Experiment B11: Concave Mirror Experiment

Purpose: To study light reflection in a concave mirror, and to determine the focal length of the mirror.

Apparatus: Concave mirror, ray box with cross wires, meter rule and screen.

Procedure:

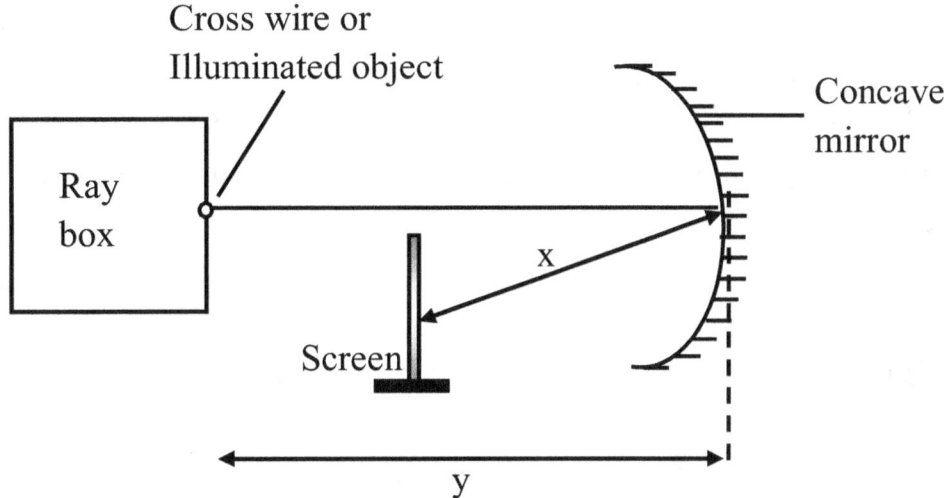

(a) Carry out the experiment by following the procedures below:

    (i) Determine and record the approximate focal length f_a of the mirror provided.

    (ii) Arrange the ray box, the mirror and the screen as shown above.

    (iii) Position the ray box in front of the concave mirror such that the cross wire (or illuminated object) is at a distance **y** = 20 cm from the concave mirror.

    (iv) Adjust the position of the screen until a sharp image of the object is formed on it. Measure and record the image distance **x**.

    (v) Evaluate $\frac{1}{x}$ and $\frac{1}{y}$.

(vi) Repeat the experiment for y = 25, 30, 35 and 40 cm respectively. In each case, determine the corresponding values of x, $\frac{1}{x}$ and $\frac{1}{y}$. Tabulate your readings.

(vii) Plot a graph of $\frac{1}{x}$ on the vertical axis against $\frac{1}{y}$ on the horizontal axis, starting both axes from the origin (0,0).

(viii) Determine the slope of the graph.

(ix) Read and record the intercept I_1 on the vertical axis and I_2 on the horizontal axis. Evaluate $K = \frac{1}{2}(I_1 + I_2)$ and $P = 1/K$.

(x) State **two** precautions taken to ensure accurate results.

(b)

    (i) What is meant by the statement *the focal length of a concave mirror is 20 cm?*

    (ii) Draw a ray diagram showing how a concave mirror is used as a shaving mirror.

Observations:

$f_a = $ _____

y (cm)	x (cm)	1/y (cm^{-1})	1/x (cm^{-1})

62

Experiment C1: Potentiometer Experiment I

Purpose: To study the relationship between the current flowing through a wire and the length of the wire.

Apparatus: Potentiometer wire, 2V accumulator, jockey, 2Ω resistor, ammeter, key and connecting wires.

Procedure:

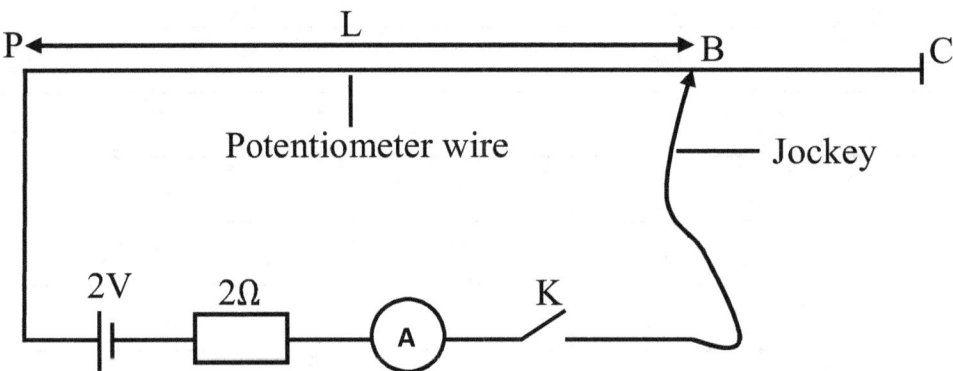

(a) Set up the circuit as shown in the diagram above.

    (i) Close the key **K**. Use the jockey to make contact with the potentiometer wire PC such that PB = L = 10 cm.

    (ii) Record the ammeter reading **I** and evaluate $\mathbf{I^{-1}}$.

    (iii) Repeat the experiment for **L** = 20, 30, 40, 50 and 60 cm respectively.

    (iv) In each case, determine and record the corresponding values of **I** and $\mathbf{I^{-1}}$. Tabulate you readings.

(v) Plot a graph of **L** on the vertical axis against **I**$^{-1}$ on the horizontal axis starting both axes from the origin (0,0).

(vi) Determine the slope **s** of the graph and intercept on the vertical axis.

(vii) Read and record the value of **I**$^{-1}$ when **L** = 0.

(viii) State **two** precautions taken to ensure accurate results.

(b)

(i) Deduce from your graph, the value of the current in the circuit when the jockey makes contact with the potentiometer wire at C.

(ii) State **two** reasons why the key in the circuit should be opened when readings are not taken.

Observations:

L (cm)	I (A)	I^{-1} (A^{-1})

_____

_____

_____

_____

Experiment C2: Ohm's Law Experiment I

Purpose: To determine the resistance of a length of wire using Ohm's law.

Apparatus: Accumulator, variable resistor, ammeter, voltmeter, key and connecting wires.

Procedure:

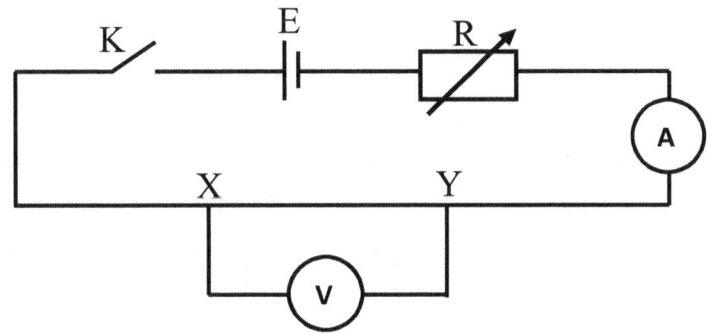

(a)

(i) Measure and record the length XY of the resistance wire provided.
(ii) Connect the circuit shown in the diagram above.

(iii) With R = 0 Ω, close the key, K. Read and record the current I_0 and the voltage drop V_0 across the resistance wire.

(iv) Setting R=1 Ω, close the key. Read and record the current, I and the corresponding voltage drop, V across the wire.

(v) Repeat the procedures for five other values of R = 5, 10, 20, 40 and 60 Ω. Tabulate your readings.

(vi) Plot a graph of V on the vertical axis against I on the horizontal axis.

(vii) Determine the slope of the graph.

(viii) State two precautions taken to ensure accurate results.

(b)

(i) Mention and state the law on which this experiment is based.
(ii) A piece of resistance wire of diameter 0.2 mm and resistance 7Ω has resistivity of $8.8 \times 10^{-7} \Omega m$, calculate the length of the wire. $[\pi = \frac{22}{7}]$.

Observations:

$I_0 =$ _____ $V_0 =$ _____

R (Ω)	I (A)	V (V)

_____

_____

_____

_____

_____

_____

_____

Experiment C3: Potentiometer Experiment II

Purpose: To study the relation between length of a wire and its resistance using a parallel combination of resistors in a potentiometer circuit.

Apparatus: Potentiometer, accumulator, 2 Ω resistor, variable resistor, a fixed resistor labeled X, key and connecting wires.

Procedure:

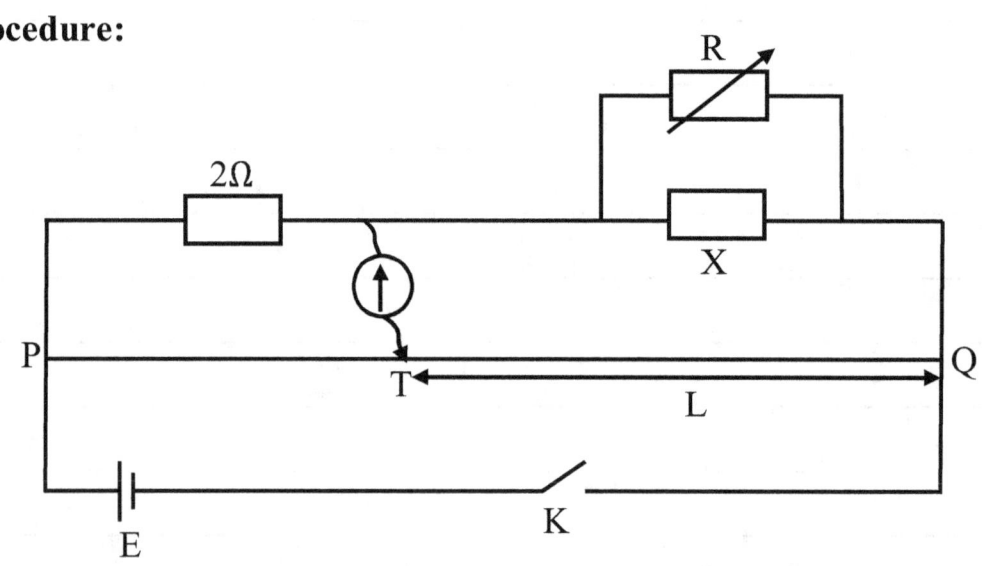

(a) Connect the circuit as shown above.

  (i) Set the value of **R** = 30 Ω. Close the key and obtain a balance at point **T** on the potentiometer wire **PQ.**

  (ii) Read and record the length **TQ = L.** Evaluate **L⁻¹** and **R⁻¹.**

  (iii) Repeat the experiment for **R** = 20, 10, 5, 3 and 1Ω respectively.

  (iv) In each case, determine and record the corresponding values of **L, L⁻¹** and **R⁻¹.**

(v) Remove the resistance box from the circuit and then determine the length L_0 corresponding to $R = 0$. Tabulate your readings.

(vi) Plot a graph of R^{-1} on the vertical axis and L^{-1} on the horizontal axis, starting both axes from the origin (0,0).

(vii) Determine the slope, **s** of the graph and its intercept, **I**, on the vertical axis.

(viii) Evaluate:

I. $K = I^{-1}$.
II. $C = L_0/s$.

(ix) State two precautions taken to ensure accurate results.

(b)

(i) Using your graph, determine the value of **L** for which $R = 15\Omega$.
(ii) If the intercept $I = 0.5 + y^{-1}$, use your graph to determine the value of **y**.

Observations:

$L_0 = $ _____

R (Ω)	L (cm)	L^{-1} (cm^{-1})	R^{-1} (Ω^{-1})

Experiment C4: Constantan Wire Experiment

Purpose: To study how the current through a piece of constantan wire changes with the length of the wire.

Apparatus: Constantan wire, crocodile clip, 2Ω resistor, accumulator ammeter, key and connecting wires.

Procedure:

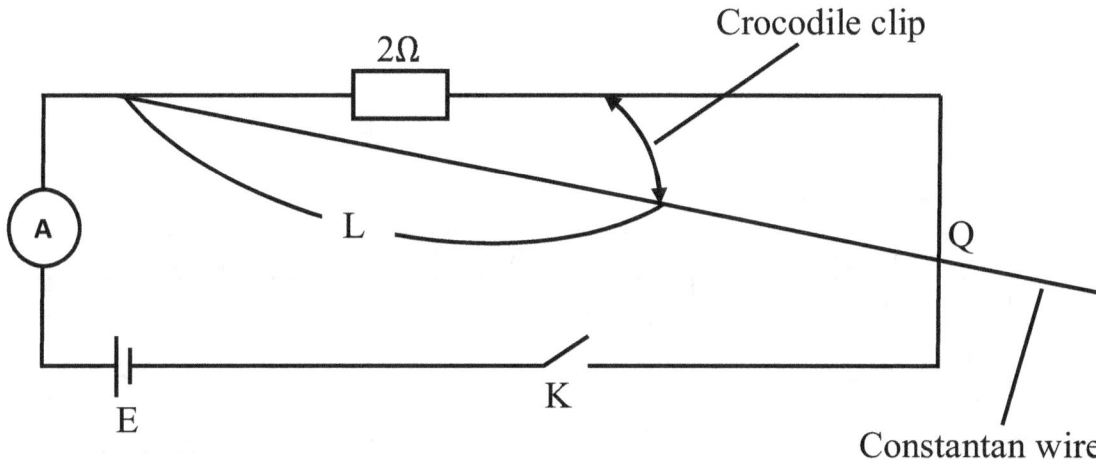

(a) You are provided with a constantan wire, a 2Ω standard resistor, an accumulator E, an ammeter A, a key K and other necessary apparatus.

    (i)   Measure and record the e.m.f of the accumulator provided.

    (ii)   Connect a circuit as shown in the diagram above.

    (iii)   Close the key, read and record the ammeter reading I_0 when the crocodile clip is not in contact with the constantan wire.

    (iv)   Open the key. With the clip making contact with the wire, when L=90cm, close the key. Read and record the ammeter reading I. Evaluate L^{-1}.

    (v)   Repeat the procedure for L = 80, 70, 60 and 50 cm.

(vi)   In each case, read and record the ammeter reading and evaluate L^{-1}. Tabulate your readings.

(vii) Plot a graph of I on the vertical axis against L^{-1} on the horizontal axis.

(viii) Determine the slope, s, of the graph, and its intercept, c, on the vertical axis.

(ix)   Evaluate $k = \dfrac{c}{s}$.

(x)   Using your graph determine the current I when L = 55cm.

(xi)   State two precautions taken to ensure accurate results.

(b)

(i) Explain what is meant by the potential difference between two points in an electric circuit.

(ii) State two factors on which the resistance of a resistance wire depends.

Observations:

$I_0 = $ _____

L (cm)	I (A)	L^{-1} (cm^{-1})

_____

_____

_____

_____

Experiment C5: Voltage-Resistance Experiment I

Purpose: To study how the voltage across a resistor changes with the resistance of the resistor.

Apparatus: Accumulator, resistance box, voltmeter, key and connecting wires.

Procedure:

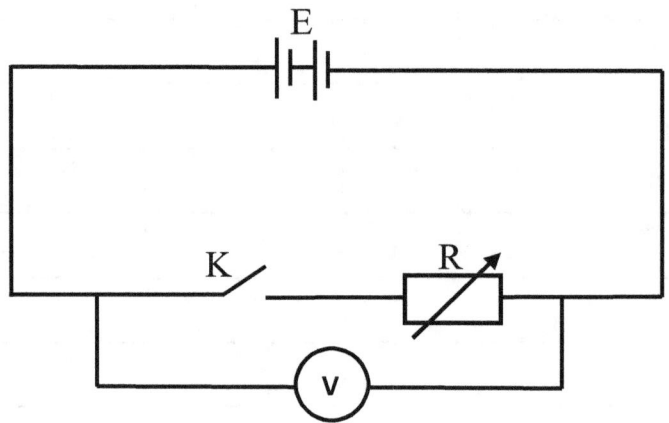

(a) Connect the circuit as shown above.

(i) Open the key **K** and record the voltmeter reading **E.**

(ii) Set the resistance **R** in the resistance box equal to 1 Ω. Close the key, read and record the potential difference **V** on the voltmeter. Evaluate R^{-1} and V^{-1}.

(iii) Repeat the experiment for **R** = 2, 3, 4, 5 and 6 Ω respectively.

(iv) In each case, read and record the voltmeter reading **V.** Also, evaluate R^{-1} and V^{-1}. Tabulate your readings.

(v) Plot a graph with V^{-1} on the vertical axis and R^{-1} on the horizontal axis, starting both axes from the origin (0,0).

(vi) Determine the slope **s** of the graph and the intercept **I** on the vertical axis.

(vii) Evaluate the expression: K = s/I.

(viii) State **two** precautions taken to ensure accurate results.

(b)

(i) If the p.d. across the resistance box in the circuit is measured. It would be observed that its value is less that the e.m.f. of the cell. Explain the reason for this difference.

(ii) Explain what is meant by the potential difference between two points in an electric circuit.

Observations:

E = _____

R (Ω)	V (V)	R⁻¹ (Ω⁻¹)	V⁻¹ (V⁻¹)

_____

_____

_____

_____

_____

Experiment C6: Potentiometer Experiment III

Purpose: To study the relationship between the length of a potentiometer wire and the voltage across it.

Apparatus: Potentiometer, accumulator, 2 standard resistors labeled R_1 and R_2, key, voltmeter, jockey and connecting wires.

Procedure:

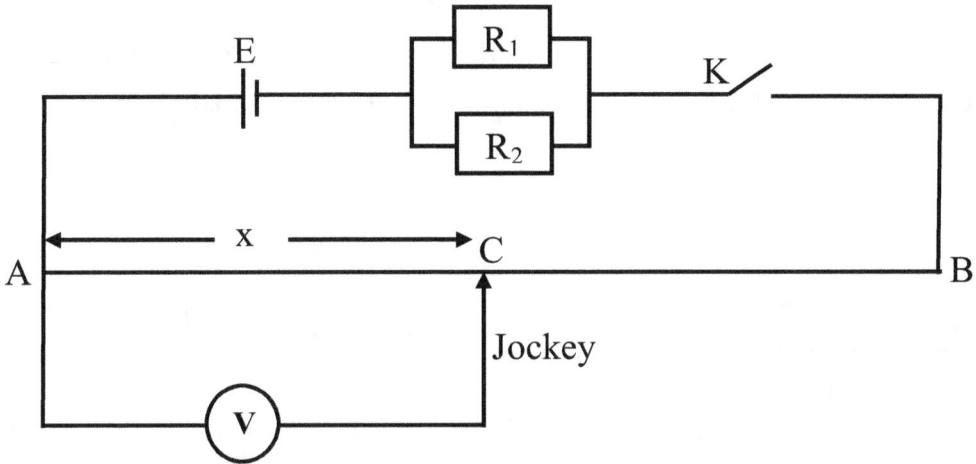

(a) You are provided with a voltmeter V, a chemical cell/battery E, two standard resistors R_1 and R_2, a potentiometer AB, a key K, a jockey and other necessary materials.

   (i) Set up a circuit as shown in the diagram above.
   (ii) Close the key K.

   (iii) Make contact with the potentiometer wire AB using the jockey at a point C such that AC = x = 20cm.

   (iv) Read and record the voltmeter reading V.

(v) Evaluate x^{-1} and V^{-1}.

(vi) Repeat the procedures for other values of x =30, 40, 50, 60 and 80cm.

(vii) Tabulate your readings.

(viii) Plot a graph with V^{-1} on the vertical axis and x^{-1} on the horizontal axis, starting both axes from the origin (0,0).

(ix) Determine the slope, s, of the graph and the intercept, c, on the vertical axis.

(x) State two precautions taken to ensure accurate results.

(b)

(i) State two devices in which Ohm's law does not apply.
(ii) A current of 1A is supplied to two resistors of resistances 2Ω and 3Ω connected in parallel. Calculate the current in each resistor.

Observations:

x (cm)	V (V)	x^{-1} (cm^{-1})	V^{-1} (V^{-1})

Experiment C7: Meter Bridge Experiment

Purpose: To study how the length of a piece of constantan wire affects its resistance using a meter bridge system.

Apparatus: Constantan wire, meter bridge, jockey, galvanometer, accumulator, 1 Ω resistor, key and connecting wires.

Procedure:

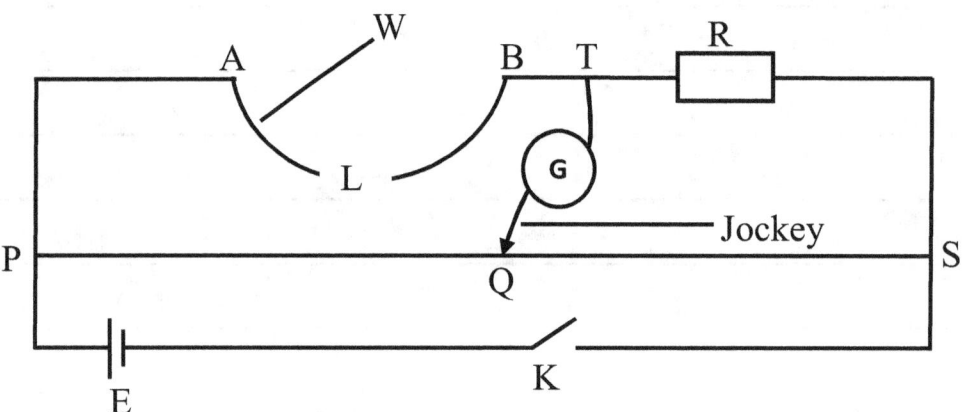

(a) You are provided with a standard resistor of resistance **R** = 1Ω and a bare constantan wire, **W.**

   (i) Connect a length **L** = 25 cm of the constantan wire, between the points **A** and **B** of the meter bridge.

   (ii) Connect other components of the circuit as shown in the diagram above.

   (iii) Now obtain a balance at **Q** on the meter bridge wire **PS** with the jockey.

   (iv) Read and record the lengths **PQ** and **QS**. Calculate the effective resistance $r = \dfrac{PQ}{QS} \times R$ of the constantan wire.

(v) Repeat the experiment for **L** = 35, 45, 55, 65 and 75 cm respectively. In each case determine the corresponding values of **PQ**, **QS** and **r**. Tabulate your readings.

(vi) Plot a graph of **r** on the vertical axis and **L** on the horizontal axis.

(vii) Determine the slope of the graph and the intercept on the vertical axis.

(b)

(i) In the circuit diagram of this experiment, **Q** represents the point of contact of the jockey with the bridge wire when the galvanometer shows null deflection. What is the potential difference between points **T** and **Q**? Explain your answer.

(ii) Two equal lengths of wire made of the same material but of different diameters have an effective resistance of 0.8 Ω when they are connected in parallel. If the cross-sectional area of one is four times the other, calculate the resistance of the thicker wire.

Observations:

L (cm)	PQ (cm)	QS (cm)	r (Ω)

(c) sciencebod

Experiment C8: Ohm's Law Experiment II

Purpose: To study the relationship between current and voltage across a potentiometer wire.

Apparatus: Ammeter, voltmeter, potentiometer wire, accumulator, a standard resistor labeled R, key, jockey and connecting wires.

Procedure:

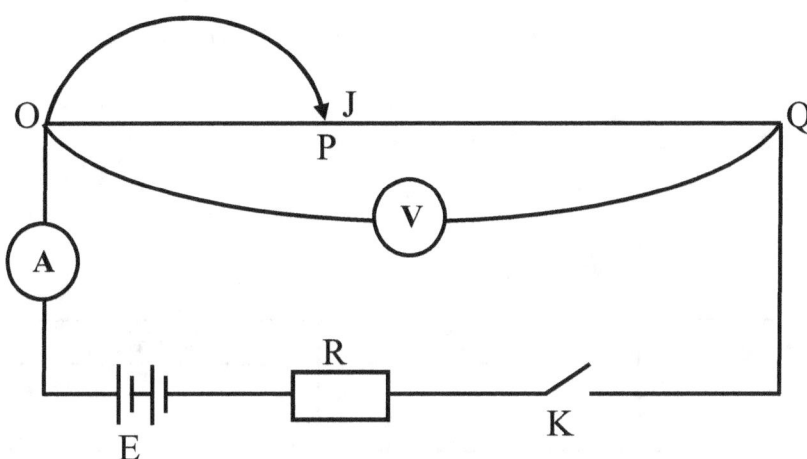

(a) Set up a circuit as illustrated in the diagram above.

(i) Close the key, K.

(ii) Read and record the ammeter reading I_0 and the voltmeter reading V_0 when jockey J is not making contact with the potentiometer wire OQ.

(iii) Using J, make a contact with the potentiometer wire OQ at a point P such that OP = 10cm.

(iv) Read and record the current I and the corresponding value of the voltage V.

(v) Repeat the procedure for other values of OP =20cm, 30cm, 40cm, 50cm and 60cm. Tabulate your readings.

(vi) Plot a graph with V on the vertical axis and I on the horizontal axis, starting both axes from the origin (0,0).

(vii) Determine the slope, s, of the graph.

(viii) Determine the value of V when $I = 0$.

(ix) State two precautions taken to ensure accurate results.

(b)

(i) State two advantages of a lead-acid accumulator over a dry leclanche cell.
(ii) A cell of e.m.f. 2V and internal resistance of 1Ω passes current through an external load of 9Ω. Calculate the potential drop across the cell.

Observations:

$I_0 =$ _____ $V_0 =$ _____

OP (cm)	I (A)	V (V)

Experiment C9: Potentiometer Experiment IV

Purpose: To study current flow in a circuit with parallel combination of a resistor and potentiometer wire.

Apparatus: Potentiometer, ammeter, resistor labeled R, accumulator, key, jockey and connecting wires.

Procedure:

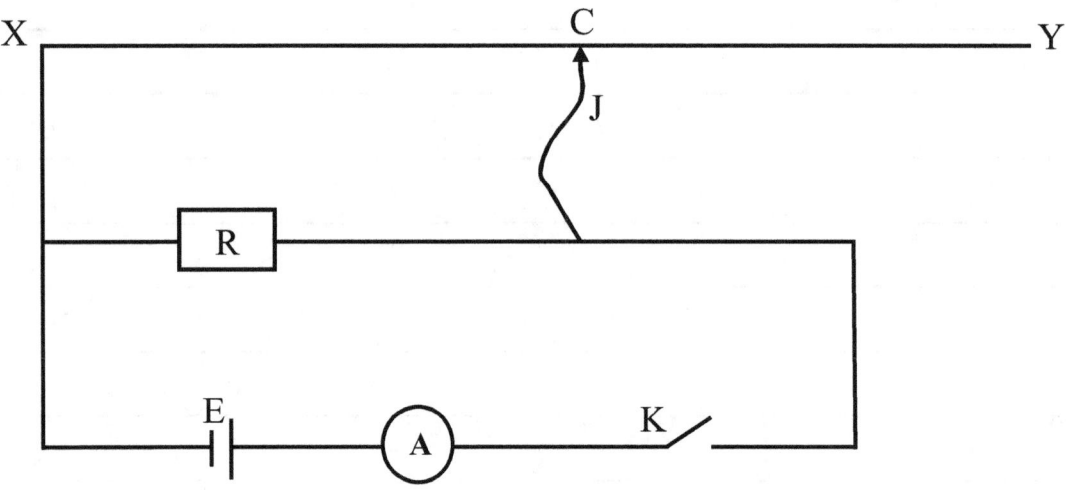

(a) You are provided with a potentiometer XY; a jockey, J; a standard resistor, R and other necessary apparatus.

    (i) Connect a circuit as shown in the diagram above.

    (ii) Close the key. Read and record the current I_o when J is not in contact with XY.

    (iii) Let J make contact with XY at C such that $XC = l = 25$cm. Close the key. Read and record the current I.

(iv) Evaluate l^{-1}.

(v) Repeat the procedure for four other values of l = 40, 55, 70 and 85cm. Tabulate your readings.

(vi) Plot a graph of I on the vertical axis against l^{-1} on the horizontal axis.

(vii) From your graph, deduce the value of I when l^{-1} = 0.

(viii) State two precautions taken to ensure accurate results.

(b)

(i) Write an expression (in terms of R_1 and R_2) for the effective resistance of two resistors, R_1 and R_2 connected in parallel.

(ii) A piece of resistance wire of diameter 0.2m and resistance 7Ω has resistivity of $8.8 \times 10^{-7}\Omega$m, calculate the length of the wire. $[\pi = \frac{22}{7}]$.

Observations:

I_0 = _____

l (cm)	I (A)	l^{-1} (cm^{-1})

_____

_____

_____

Experiment C10: Voltage-Resistance Experiment II

Purpose: To study how resistors in a circuit share the voltage across the circuit.

Apparatus: Two resistance boxes, a standard resistor labeled R_x, two keys, a voltmeter, an accumulator and connecting wires.

Procedure:

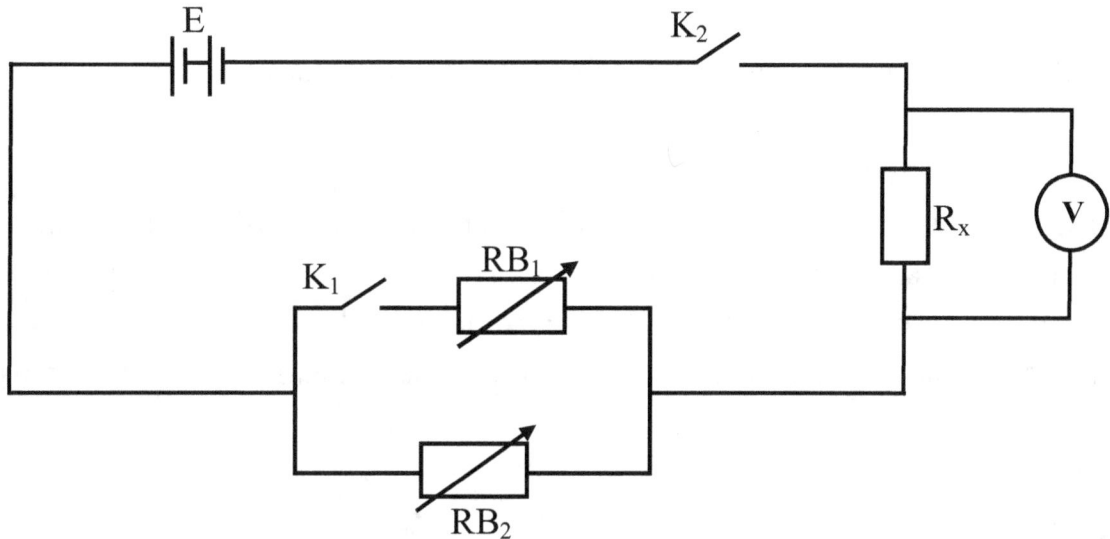

(a) You have been provided with an accumulator E, a standard resistor R_x, two resistance boxes RB_1 and RB_2, two keys K_1 and K_2, and other necessary apparatus.

   (i) Measure and record the e.m.f. of the accumulator.
   (ii) Connect a circuit as shown above.

   (iii) Set the resistance R, in the resistance boxes such that R in RB_1 = R in RB_2 = 1Ω.

   (iv) With K_1 open and K_2 closed, measure and record the potential difference V_o across the standard resistor R_x.

(v) Close K_1 and K_2. Read and record the potential difference V_1 across R_x.

(vi) Evaluate V_1^{-1}.

(vii) Repeat procedure (v) for four other values of R = 2, 3, 4, and 5Ω respectively. In each case, ensure that the value of R in RB_1 is equal to the value of R in RB_2.

(viii) Evaluate V_1^{-1} in each case. Tabulate your readings.

(ix) Plot a graph of V_1^{-1} on the vertical axis against R on the horizontal axis, starting both axes from the origin (0,0).

(x) Determine the slope, s, of the graph and the intercept, I, on the vertical axis.

(xi) Evaluate $y = \dfrac{I}{s}$.

(xii) State two precautions taken to ensure accurate results.

(b)

(i) Explain why voltmeters are usually connected in parallel to a resistor while ammeters are connected in series to a resistor when respectively reading the voltage across the resistor and the current through it.

(ii) Determine the effective resistance of six 5-Ω resistors connected in parallel in a circuit.

Observations:

E.m.f of accumulator = _____ V_o = _____

R (Ω)	V_1 (V)	V_1^{-1} (V^{-1})

_____

_____

_____

_____

_____

_____

_____

_____

_____

_____

_____

_____

_____

_____

_____

_____

www.ingramcontent.com/pod-product-compliance
Lightning Source LLC
Chambersburg PA
CBHW081452170526
45166CB00008B/2402